CAREER IN ENGINEERING

A Guide for Engineering Students, Interns, and Professionals

Calvine Odero

Copyright © 2022 Calvine Odhiambo odero

All rights reserved

The characters and events portrayed in this book are fictitious. Any similarity to real persons, living or dead, is coincidental and not intended by the author.

No part of this book may be reproduced, or stored in a retrieval system, or transmitted in any form or by any means, electronic, mechanical, photocopying, recording, or otherwise, without express written permission of the publisher.

I dedicate this book to students, interns, and professionals in the field of Engineering.

"The ideal engineer is a composite ... He is not a scientist, he is not a mathematician, he is not a sociologist or a writer; but he may use the knowledge and techniques of any or all of these disciplines in solving engineering problems."

NATHAN W. DOUGHERTY, AMERICAN CIVIL ENGINEER

CONTENTS

Title Page
Copyright
Dedication
Epigraph
ELECTRICAL AND ELECTRONICS ENGINEER   1
CIVIL ENGINEER   3
MECHANICAL ENGINEER   5
COMPUTER ENGINEER   7
TELECOMMUNICATIONS ENGINEER   9
CHEMICAL ENGINEER   11
INDUSTRIAL ENGINEER   13
NUCLEAR ENGINEER   15
AERONAUTICAL ENGINEER   17
ARCHITECTURAL ENGINEER   19
REFERENCE   21
About The Author   23

ELECTRICAL AND ELECTRONICS ENGINEER

1. Job Summary

Designing, developing, and testing electrical and electronic devices and systems.

2. Duties and Responsibilities

Apply electrical power to improve products, test electrical equipment to ensure quality specifications, innovate electrical components for commercial applications, inspect electrical and electronic instruments to ensure safety standards, and investigate electrical and electronic complaints and recommend solutions.

3. Authority and Reporting Relationship

The engineer leads the electrical and electronic innovation project and is accountable to the Head of the Engineering Department.

4. Education

Bachelor of Science in Electrical and Electronics Engineering

5. Professional Membership

Member of the Institute of Electrical and Electronics Engineers (IEEE)

6. Professional Certification

Certified Systems Engineer Professional (CSEP)

7. Key Skills

Leadership, research, creativity and innovation, teamwork, and communication.

8. Work Experience

Relevant electrical and electronic work experience of 3 years.

9. Median Annual Pay

US$ 101,780 (U.S. Bureau of Labor Statistics, 2022)

10. Challenge and Opportunity

Electrical and Electronic Engineers must pursue continuous education, training, and professional development.

Organizations have sufficient opportunities for Electrical and Electronic Engineers to rise to senior management and leadership positions in energy projects and programs.

CIVIL ENGINEER

1. Job Summary

Design, develop, test, supervise, and maintain infrastructure systems and projects.

2. Duties and Responsibilities

Analyze survey reports to plan infrastructure projects, participate in making the project budget, compile permit applications to comply with state and federal government regulations, perform soil testing to determine the strength of foundations, and maintain organizational infrastructure.

3. Authority and Reporting Relationship

Leader of the infrastructure project, and is accountable to the Head of the Engineering Department.

4. Education

Bachelor of Science in Civil Engineering

5. Professional Membership

Member of the American Society of Civil Engineers (ASCE)

6. Professional Certifications

Project Management Professional (PMP)

7. Key Skills

Teamwork, research, innovation and creativity, communication, and problem-solving.

8. Work Experience

Relevant work experience of 5 years in the construction environment.

9. Median Annual Pay

US$ 88,050 (U.S. Bureau of Labor Statistics, 2022)

10. Challenge and Opportunity

Most construction projects are complex, long-term, and costly to implement.

A good opportunity is a high and competitive demand for civil engineers in the area of sustainable infrastructure engineering.

MECHANICAL ENGINEER

1. Job Summary

Research, develop, and test mechanical devices, which involve engines, thermal sensors, and machines.

2. Duties and Responsibilities

Analyze issues and determine how thermal or mechanical systems and devices may offer solutions, design and develop prototypes of thermal and mechanical equipment, supervise the production of the devices, conduct research to improve the performance of the mechanical and thermal products, and implement robotic sensors.

3. Authority and Reporting Relationship

In charge of the mechanical and thermal projects in the organization, and is accountable to the Production Manager.

4. Education

Bachelor of Science in Mechanical Engineering

5. Professional Membership

Member of the Institution of Mechanical Engineers

6. Professional Certification

Certified Systems Engineering Professional (CSEP)

7. Key Skills

Communication, research, analytical, problem-solving, and innovation.

8. Work Experience

Experience of at least 5 years in a mechanical engineering work environment.

9. Median Annual Pay

US$ 95,300 (U.S. Bureau of Labor Statistics, 2022)

10. Challenge and Opportunity

Continuous training on emerging innovations in the field of mechanical engineering.

Sufficient career progression opportunities in project and program management.

COMPUTER ENGINEER

1. Job Summary

Research, design, and test Information and Communication Technology (ICT) systems and components.

2. Duties and Responsibilities

Develop computer hardware, test the design of the ICT hardware, modify the design of the hardware accordingly, upgrade existing computer devices in line with new software, and supervise the production of computer hardware components and systems.

3. Authority and Reporting Relationship

Lead the computer hardware innovation team, and is accountable to the ICT Manager in the organization.

4. Education

Bachelor of Science in Computer Engineering

5. Professional Membership

Member of the Association for Computing Machinery (ACM)

6. Professional Certification

Certified Hardware and Technology Engineer (CHTE)

7. Key Skills

Project management, research, creativity and innovation, analytical, and communication.

8. Work Experience

Work experience of 5 years in computer hardware design, development, and test work environment.

9. Median Annual Pay

US$ 128,170 (U.S. Bureau of Labor Statistics, 2022)

10. Challenge and Opportunity

Continuous training and certifications in the dynamic and innovative ICT work environment.

ICT innovations offer a good opportunity to develop high-quality computer hardware devices and systems.

TELECOMMUNICATIONS ENGINEER

1. Job Summary

Design, configure, and test voice, video, and data systems of communication.

2. Duties and Responsibilities

Installation of networks and equipment of communication, maintenance of the communication systems, upgrading the communication devices and programs, ensuring compatibility and reliability of equipment and networks, and supervising technicians in the telecommunication project team.

3. Authority and Reporting Relationship

Leader of the telecommunication team, and accountable to the Head of the Engineering department.

4. Education

Bachelor of Science in Telecommunications Engineering

5. Professional Membership

Member of the Association of Communication Engineers (ACE)

6. Professional Certification

Certified Telecommunications Network Specialist (CTNS)

7. Key Skills

Creativity and innovation, communication, research, problem-solving, and teamwork.

8. Work Experience

Work experience of 5 years in designing and maintaining telecommunication systems.

9. Median Annual Pay

US$ 88,174 (U.S. Bureau of Labor Statistics, 2022)

10. Challenge and Opportunity

The telecommunication engineer must improve competencies continuously due to regular innovations in the telecommunications industry.

Opportunity for career growth is high in senior management and leadership roles of projects and organizations.

CHEMICAL ENGINEER

1. Job Summary

Applies scientific, mathematical, and technological concepts to improve the production of foods, drugs, beverages, and fuel products.

2. Duties and Responsibilities

Implement research and innovations that enhance the quality and quantity of chemical products, implement safety procedures in the use of chemicals, develop processes that separate liquids and gases, ensure that production processes are environmentally sustainable, and implement oxidation and polymerization processes to create different chemical and plastic products.

3. Authority and Reporting Relationship

Leader of the chemical product development team, and is accountable to the Production Manager.

4. Education

Bachelor of Science in Chemical Engineering

5. Professional Membership

Member of the American Chemical Society (ACS)

6. Professional Certification

Certified Chemical Engineer

7. Key Skills

Team leadership, communication, creativity and innovation, research, and problem-solving.

8. Work Experience

Experience of 5 years in a chemical production work environment.

9. Average Annual Salary

US$ 105,550 (U.S. Bureau of Labor Statistics, 2022)

10. Challenge and Opportunity

The work environment is hazardous and requires a high level of health and safety policies and programs.

The chemical engineering work environment offers sufficient opportunities to innovate new products, which have superior quality.

INDUSTRIAL ENGINEER

1. Job Summary

Management of an effective and efficient system that integrates staff, machines, data, and energy to produce high-quality products and services.

2. Duties and Responsibilities

Apply the production specifications and schedules, ensure maximum efficiency in production processes, use the management control system effectively, prepare cost analysis for the production processes, and address issues or complaints from the clients.

3. Authority and Reporting Relationship

Supervise the staff in the production environment, and is accountable to the Production Manager.

4. Education

Bachelor of Science in Industrial Engineering

5. Professional Membership

Members of the Institute of Industrial and Systems Engineers (IISE)

6. Professional Certification

Certified Manufacturing Engineer

7. Key Skills

Creativity and innovation, critical thinking, problem-solving, team leadership, and communication.

8. Work Experience

Experience of 5 years in a production or manufacturing environment.

9. Median Annual Pay

US$ 95,300 (U.S. Bureau of Labor Statistics, 2022)

10. Challenge and Opportunity

High-pressure work environment as the quality and output specifications must be achieved.

The high demand for industrial engineers enhances the competitiveness of the career in senior management and leadership roles.

NUCLEAR ENGINEER

1. Job Summary

Research, design, and develop systems and instruments that generate nuclear benefits in the areas of energy and radiation.

2. Duties and Responsibilities

Develop devices that use nuclear reactor cores and radiation shields, maintain the nuclear plant, observe health and safety standards in the workplace, develop operational plans and instructions applicable in nuclear operations, and ensure proper disposal of wastes from the nuclear processes.

3. Authority and Reporting Relationship

Head of the nuclear energy team, and is accountable to the Head of the Engineering department.

4. Education

Bachelor of Science in Nuclear Engineering

5. Professional Membership

Member of the American Nuclear Society (ANS)

6. Professional Certification

Professional Engineer (PE) in Nuclear Engineering

7. Key Skills

Creativity and innovation, project management, communication, team leadership, and problem-solving.

8. Work Experience

Work experience of 5 years in a nuclear energy application

environment.

9. Median Annual Pay

US$ 120,380 (U.S. Bureau of Labor Statistics, 2022)

10. Challenge and Opportunity

The job has high-stress levels as it requires high accuracy and no room for mistakes.

The job holder has the opportunity to innovate valuable nuclear energy solutions that enhance the quality of life of humanity.

AERONAUTICAL ENGINEER

1. Job Summary

Design, manufacture, and test aircraft and aerospace equipment.

2. Duties and Responsibilities

Coordinate the production of aircraft, ensure that aircraft projects are financially feasible, ensure safety in the operation of aircraft equipment, ensure product design satisfies engineering principles, and satisfy the requirements and interests of clients.

3. Authority and Reporting Relationship

Leads the aircraft production team, and is accountable to the Airport Operations Manager.

4. Education

Bachelor of Education in Aeronautical Engineering

5. Professional Membership

Member of the American Institute of Aeronautics and Astronautics (AIAA)

6. Professional Certification

Aircraft Stability and Control Certification

7. Key Skills

Team leadership, creativity and innovation, research, critical thinking, and problem-solving.

8. Work Experience

Experience of 5 years in aircraft production and maintenance environment.

9. Average Annual Salary

US$ 122,270 (U.S. Bureau of Labor Statistics, 2022)

10. Challenge and Opportunity

The workload for the job is high as aircraft design, maintenance, and testing is done daily.

A significant opportunity for the job holder is the creativity and innovation that facilitates the creation of new and superior technology in aircraft operations.

ARCHITECTURAL ENGINEER

1. Job Summary

Plan and design buildings and related physical structures.

2. Duties and Responsibilities

Understand the requirements of the client concerning the structure, estimate the cost of construction, develop structure specifications, scale drawings using computer software, and write contract documents for contractors.

3. Authority and Reporting Relationship

Lead the construction design team, and is accountable to the Engineering Manager.

4. Education

Bachelor of Science in Architectural Engineering

5. Professional Membership

Member of the Architectural Engineering Institute (AEI)

6. Professional Certification

Designated Design-Build Professional Certification

7. Key Skills

Team leadership, communication, project management, creativity and innovation, and ICT.

8. Work Experience

Experience of 5 years in the design and supervision of building projects.

9. Median Annual Pay

US$ 80,180 (U.S. Bureau of Labor Statistics, 2022)

10. Challenge and Opportunity

Architecture is a high-pressure work as it is responsible for the outcomes of built environments.

Job responsibilities are diverse and the remuneration is attractive.

REFERENCE

U.S. Bureau of Labor Statistics. (2022). Occupational Outlook Handbook. Retrieved from https://www.bls.gov/ooh/

ABOUT THE AUTHOR

Calvine Odhiambo Odero

A Human Resource Management (HRM) professional, currently working as a Senior Human Resource Management Officer.

www.ingramcontent.com/pod-product-compliance
Lightning Source LLC
Chambersburg PA
CBHW050327220526
45465CB00005B/2162